MIDWAY JR. HIGH LIBRARY
MENAN, IDAHO 83434

# Television and Video

**HELEN MINTERN**

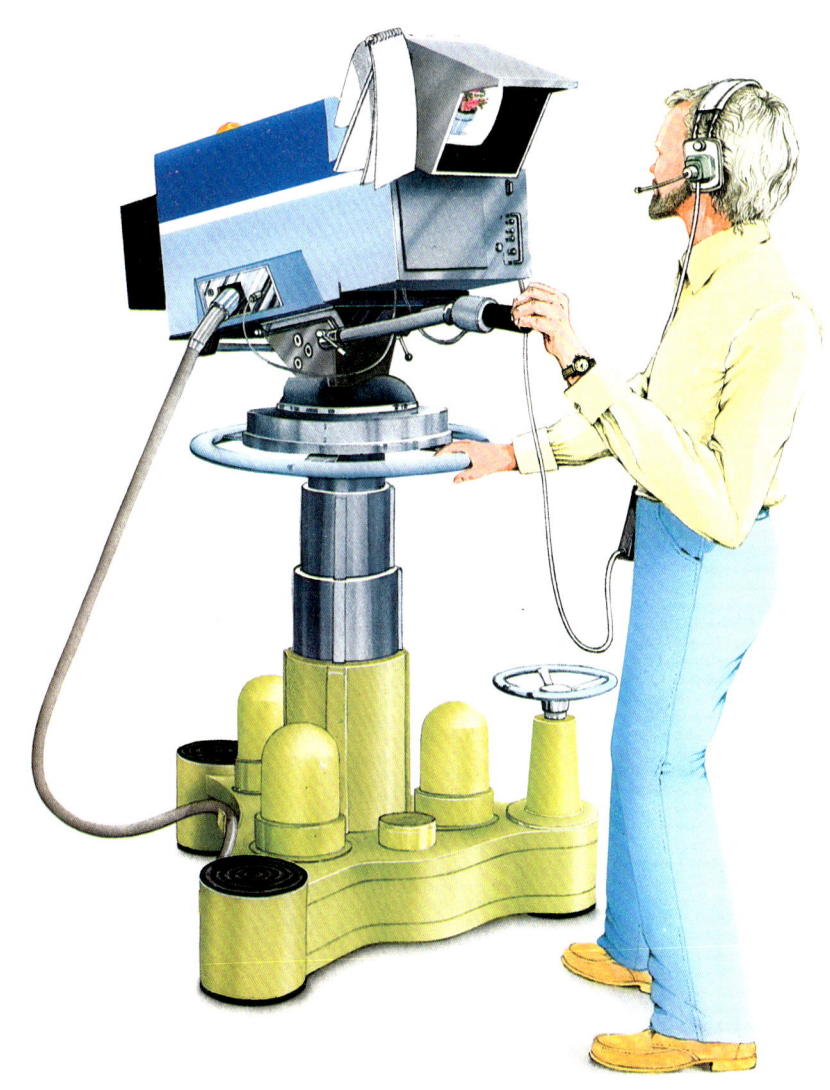

**WARWICK PRESS**

Published 1983 by Warwick Press,
387 Park Avenue South, New York, New York 10016.

First published in Great Britain by Kingfisher
Books Limited, 1983.

Copyright © Grisewood & Dempsey Ltd 1983.

Printed in Italy by Vallardi Industrie Grafiche, Milan.

6 5 4 3 2 1   All rights reserved

ISBN 0-531-09225-9
Library of Congress Catalog Card No. 83-50344

**Artists**
Dave Etchell and John Ridyard
Bill Le Fever/Linda Rogers Associates
Maltings Partnership
Janos Marffy/Jillian Burgess
Mike Saunders/Jillian Burgess

# Contents

| | |
|---|---|
| The Scope of TV and Video | 8 |
| The TV Receiver | 10 |
| Pictures from the Air | 12 |
| On the Studio Floor | 14 |
| In the Gallery | 16 |
| Outside Broadcasts | 18 |
| Film and Video Tape | 20 |
| Satellites and Cable | 22 |
| Developments in TV | 24 |
| Video Recorders and Video Disks | 26 |
| Making a Video: Equipment | 28 |
| Making a Video: Shooting | 30 |
| Video at Work | 32 |
| Glossary | 34 |
| Index | 36 |

# The Scope of TV and Video

When you are grown-up and your children are going to school, this book may not exist. In fact, schools as you know them may not exist either, and libraries with books may be museums. All this will happen because of television and video. Television was invented during the 1920s by John Logie Baird.

**Studying by Television**

Let's visit a home of the future, say in the year 2000, and see what everyone is doing. Alice is 12 years old. She is not wasting time watching television; she is at school. That's her teacher on the screen. She manages to see Alice once a week to check her written work, but not for long. By teaching on television she could have a thousand pupils in her class at once, but she doesn't have more than a hundred. Alice likes to "go to school" in the living room where there is a row of flat screens against the wall. She wears headphones to listen to her teacher.

Alice's brother, Peter, likes to work by himself in his bedroom with a smaller, personal screen. He is 20 years old and, although he lives in California, he's studying with the Massachusetts Institute of Technology. His microcomputer and screen are linked by telephone to the local library. They are sending Peter a new article written in Cambridge. They received it overnight from Massachusetts during the cheaper-rate computer time.

**Work and Leisure**

Father has worked at home for the last five years, ever since his supermarket became fully automated. As supply manager for the supermarket, he checks the stock on the shelves every Monday morning visually through the closed circuit television cameras. On his home screen he can also study the computer totals produced by the automatic checkout tills.

Mom is watching a live television program. Her favorite daytime program is the 24-hour European news station which the family receive through their satellite dish receiver on the roof.

Grandad Jones is the only member of the family who uses the video disk. At the moment he's looking at a dahlia catalog, and the video disk gives the best picture available on any system.

Grandmother Jones can hardly walk and spends her time watching the goings-on out in the street through the local closed circuit camera system. The council originally set up the system to help stop burglaries.

Alice

**Television and Video in the Year 2000**

# The TV Receiver

Whatever else you may have in your home in the year 2000, you can be sure there will be some kind of television set. The word *television* actually means "pictures from far away", but to understand how the set produces these pictures it might be better to use the word "dottyvision" instead.

## How Television Works

Your television set has three main parts: the tuner, the loudspeaker and the cathode ray tube, as well as various electronic parts which deal with the signal for the cathode ray tube.

The tuner receives different signals from several television stations through the antenna, which picks the signals out of the air. The tuner separates the pictures from the sound. These have traveled together from the transmitter (see pages 12–13). The part of the tuner you can see is a set of knobs or buttons which enable you to tune in to each channel.

The loudspeaker receives the sound as waves of electronic signals. The signals activate a magnet in the loudspeaker. In turn, the magnet makes a cardboard cone vibrate many times a second, disturbing the air and producing the noises we hear.

## Receiving Color Pictures

The picture signal is sent to the cathode ray tube. You will recognize the wide end of this as the television screen. The screen is coated inside with chemical phosphor dots. These are arranged in sets of three to produce the three different colors of television – red, blue and green. Behind the screen is a thin plate called a mask with vertical slots cut in it. Each slot is for one group of dots.

The narrow end of the cathode ray tube is hidden by the casing of the set. It contains three electron guns. The guns spray electrons on to the phosphor dots through the slots in the mask. The phosphor dots glow with color to form a pattern, or picture, when hit by the electrons.

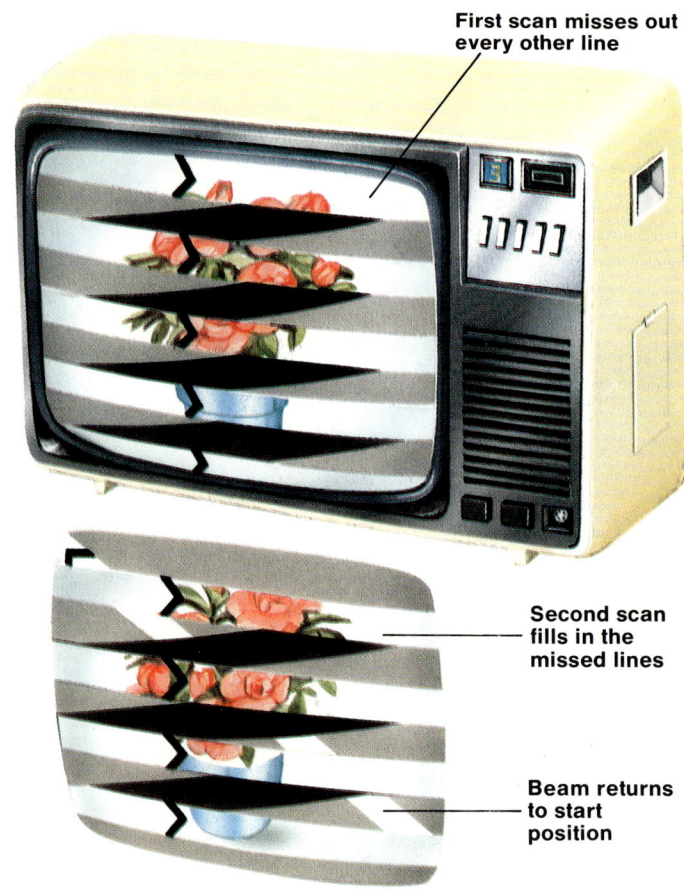

First scan misses out every other line

Second scan fills in the missed lines

Beam returns to start position

▲ The picture looks steady, but the electron beams scan the screen 50 times each second. The phosphor dots glow red, blue or green to produce the pattern of each picture. If you go very close to the screen you can see the dots.

▶ The main part of the TV set is the cathode ray tube. The big end is the screen you see. The narrow end contains three electron guns. These spray electrons through the shadow mask on to the phosphor dots. Some sets use only one gun.

Each dot can only receive electrons from one gun because of the way the slots are positioned. Some of the newest sets have only one gun to spray all three colors.

The spraying action is like a garden sprinkler, zig-zagging horizontally across the screen in lines. In Europe, television pictures have 625 lines and in the United States they have 525. The guns carry out two scans. The first misses out every other line. The second fills in the

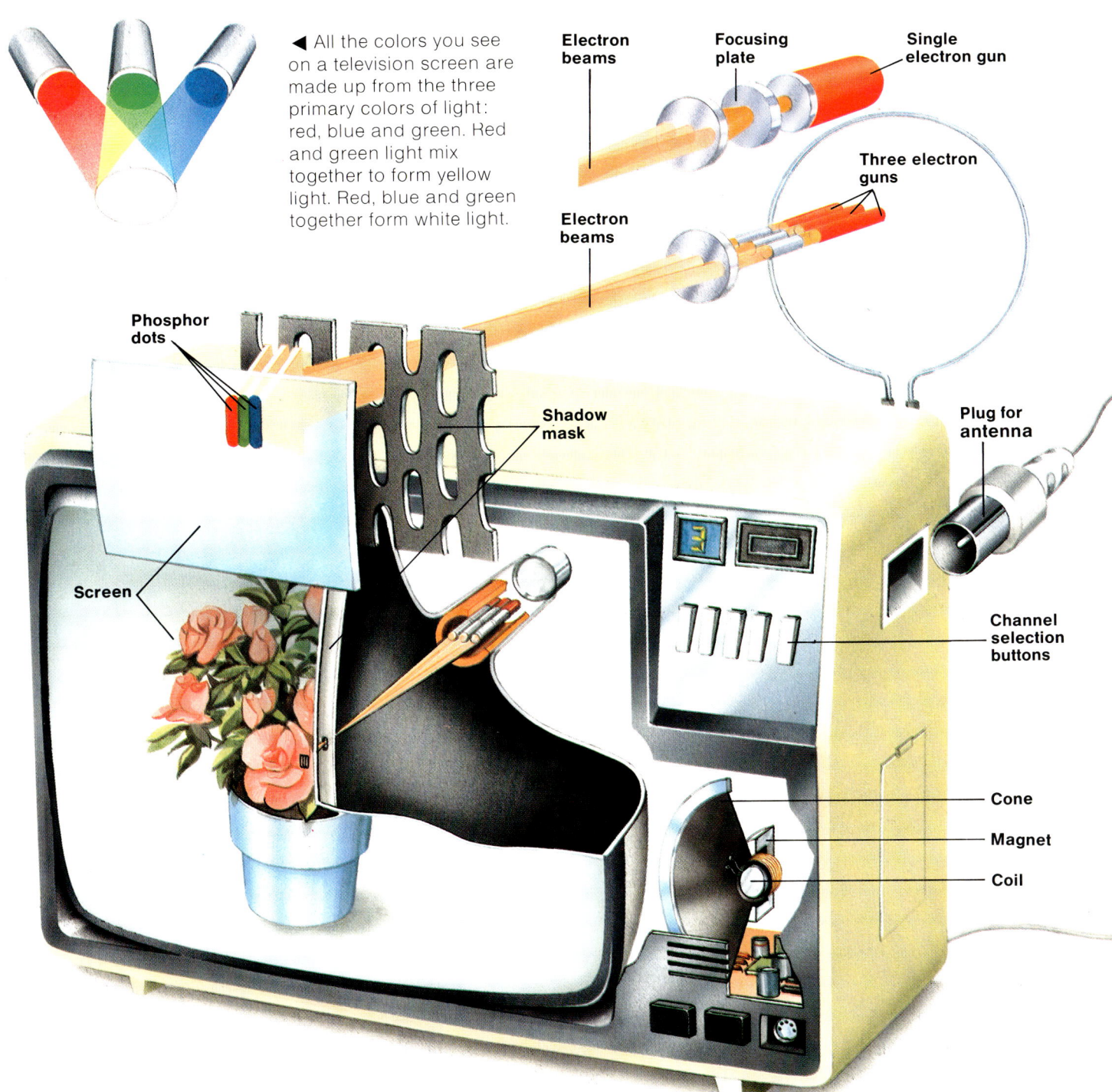

All the colors you see on a television screen are made up from the three primary colors of light: red, blue and green. Red and green light mix together to form yellow light. Red, blue and green together form white light.

missed lines. All this happens 50 times every second.

## Black and White Pictures

The same phosphor dots and electron guns are used to produce black and white pictures. The three colors are mixed to make black, white and gray shades. It may be difficult to understand how this is possible because we are used to mixing paint colors, which behave differently from the colors in light.

If you want to experiment with mixing light, try this. Take three torches and put red film in front of one, blue film in front of another, and green film in front of the third. Now shine the red, green and blue lights on to a white screen. The area where all three colors overlap will be white.

# Pictures from the Air

Like the simplest pocket camera, the television camera has an "eye", or lens. Light from objects enters the camera through the lens and is changed into electrical signals. These are combined with carrier waves and sent out into the air from transmitters. The television set at home picks up these signals and converts them into a picture on the screen.

## How the Camera Works

Inside the camera are three tubes, and at the end of each tube is a target plate. These plates are coated with a light-sensitive material which alters when touched by light. All objects reflect light. Light coming from the subject enters through the lens of the camera. It touches the target plates and the light-sensitive material releases electrons which move down the tube.

An electron is part of an atom. When electrons start breaking away from the atoms on the target plate, they leave behind a pattern of positive electrical charges. These charges contain all the information needed to produce a picture on a television screen.

Once this pattern of information has been created, it needs to be carried out of the camera. This is done by using electron guns, which are situated at the other end of the three tubes. Each gun fires electrons at its target plate in order to read, or scan, the pattern and turn it into electrical signals. The burst of electrons scans the target plate in a similar way to the electron guns inside the television set, and is also very fast. The target is scanned 50 times every second.

## Color Television

Before the arrival of color television, the camera had only one tube. Now that colors are being transmitted, light coming through the camera lens has to be split up into its three primary colors: red, blue and green.

Each of the three camera tubes handles one color. Light that looks white to us is actually

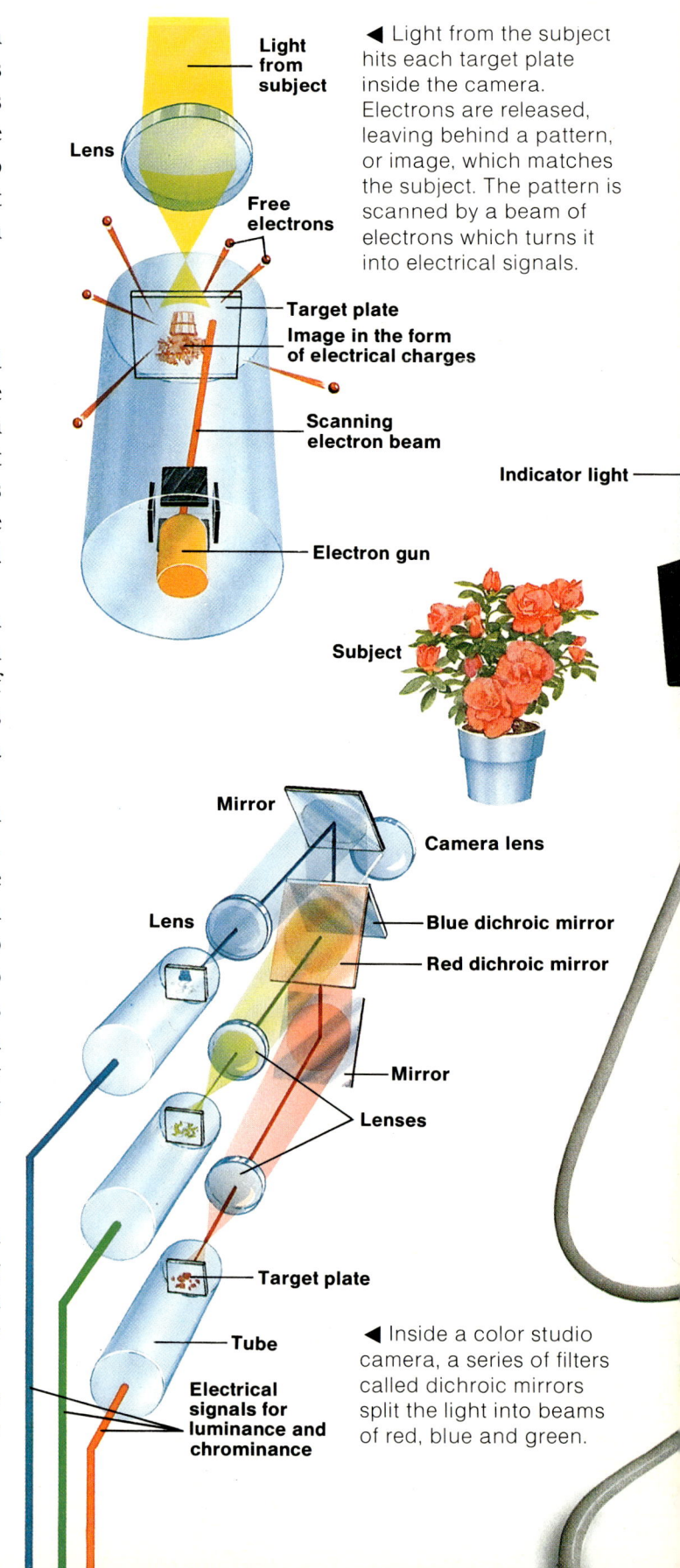

◀ Light from the subject hits each target plate inside the camera. Electrons are released, leaving behind a pattern, or image, which matches the subject. The pattern is scanned by a beam of electrons which turns it into electrical signals.

◀ Inside a color studio camera, a series of filters called dichroic mirrors split the light into beams of red, blue and green.

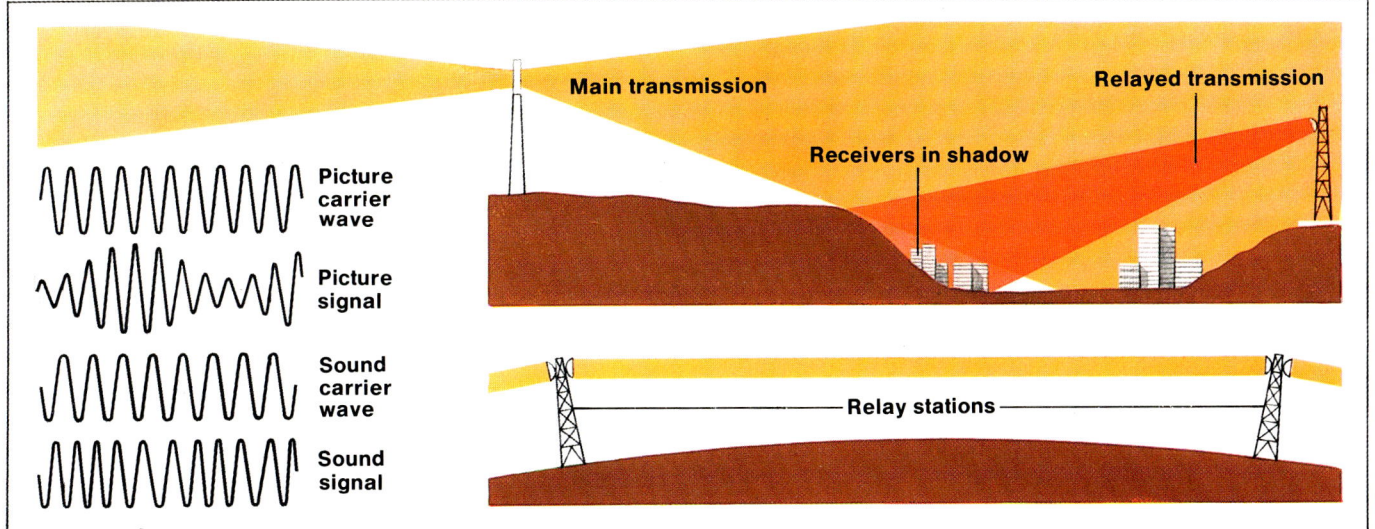

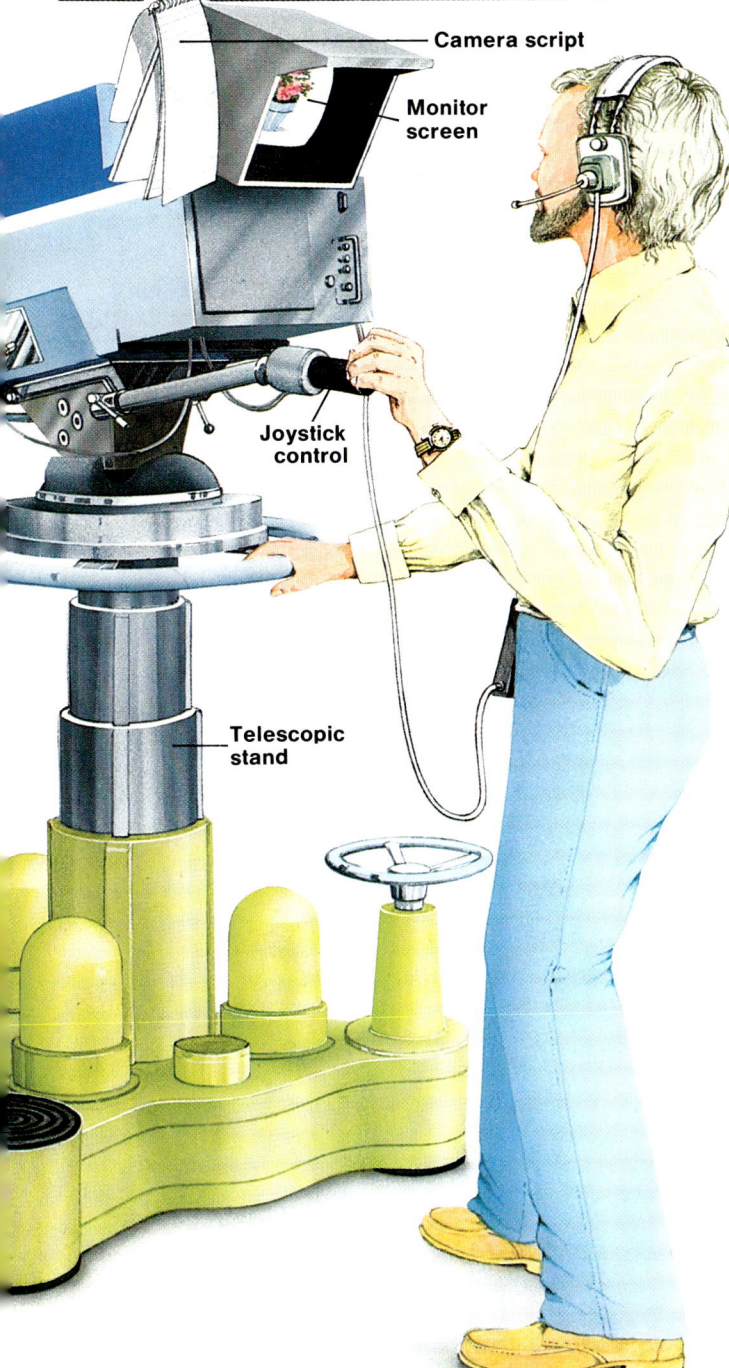

◀ The studio camera is big and heavy but can be moved easily. The cameraman can slide it backward, forward and sideways (tracking). He controls the lens from the joystick, zooming closer to the subject or pulling back, without moving.

▲ At the transmitter, the engineers use two oscillators to produce carrier waves. One wave is mixed with the picture signal, the other with the sound signal. These are strengthened (amplified) and mixed together before being sent out into the air.

made up from all the colors of the rainbow. In the color television camera, there is a series of filters called dichroic mirrors which split the light into beams of red, blue and green light. The beams are transformed into electrical signals.

## Transmitting the Signals

The three tubes first mix the light to produce a signal called the luminance signal. This carries only light and shade and can also be received by black-and-white sets. Then the tubes produce the color or chrominance signal. Both signals are sent to the transmitter.

At the transmitter, the engineers create two radio waves called carrier waves. The sound signal which matches the picture is carried by one wave. The picture signals (luminance and chrominance) are carried by the other.

Finally, the two carrier waves themselves are combined and the signal is sent out. The carrier waves can't travel through mountains or buildings, and they need strengthening every 60 miles (97 kilometers). So they are sent across a network of transmitters and booster stations.

# On the Studio Floor

The picture you see on television at home is only a tiny part of what's going on in the studio. Pull the camera back to "widen the shot" and you will see much more.

### The Cameramen

First you will see four or five cameras worked by people wearing headphones. In spite of their size, television cameras can glide like swans and move at the touch of a fingertip. Sometimes a camera is mounted to swing in the air for spectacular shots.

Each cameraman has a small monitor on his camera to show him what's being transmitted to the viewers, as well as a viewfinder to see his own picture. He can change shot by moving the lens in and out (zooming) and by moving the camera along the floor (tracking).

### The Director and Floor Manager

The director gives all the orders in the studio. He or she gives instructions through a microphone and uses special words to say what kind of shot is required. For example, if the director asks for a "BCU" (big close-up) of someone, he or she wants the cameraman to show the person's face only.

The person rushing around the studio wearing headphones is the floor manager who organizes everything taking place on the studio floor. He or she takes care of the program guests and generally makes sure that everything runs smoothly. The studio itself, especially if there is no audience, is divided into areas so that the studio staff can move from one item to another without stopping.

The director of a children's magazine program might divide the studio into three separate areas. The smallest may have a desk for a newsreader or frontman to introduce items. Another might have some screens, called "flats", to display material such as paintings, and the third area, a large space near the studio doors, might be used to stage big items.

### Sound and Light

Sound is just as important as pictures. Well before the program begins, the sound people have been down on the floor fixing microphones, or "mikes", to the desk. If the mike swings overhead it is known as a boom.

Lights are positioned on the studio ceiling. These have shutters, or "barn doors", in front of them to control the amount of light needed. The electricians (called "sparks") adjust them with long poles. The entire set is controlled by a lighting panel so that different effects and colors can be created for different items.

Waiting well "out of shot" are the scenery people who move large items like the flats, and the property people whose job is to move small things such as the newsreader's chair.

A flashing red light in the corner of the studio shows that there's just one minute before the program goes on the air. It is time to move upstairs into the gallery, or control room. From now on, the director controls the studio through the floor manager, who receives instructions through headphones.

▲ Cameramen wear headphones so that they can receive instructions from the director in the control room.

▶ Shooting drama in the studio. Each camera gives a different view of the set. The microphones and lights are out of shot.

# In the Gallery

In the final seconds before a program begins, the director may wish everyone "Good luck!" over the microphone. So many things can go wrong and everyone has to concentrate very hard all the time.

**Who's Who in the Gallery**
The director sits in front of a control panel called a mixing desk. On one side of the director is the vision mixer who "cuts up", or produces, the picture the director asks for by operating control keys and switches.

On the other side of the director is a production assistant who constantly tells everyone, including the frontman, how many seconds there are left on an item. The frontman listens through an earpiece invisible to the viewers. He or she can then read at the correct speed to fit in with the recorded material.

At least two other people sit at the control desk, although it isn't so obvious what they do. They are in fact the first and second engineers, who are in charge of everything – from lighting to the technical quality of pictures being sent to the gallery from outside the studio.

**The Monitors**
Above the control desk are lines of screens called monitors. These are not television sets, but closed circuit screens which show the

◀ The gallery, or control room. The director sits in front of the control desk, and uses a microphone to talk to the floor manager and cameramen on the studio floor. The bottom row of monitors shows the picture from each camera. The vision mixer cuts to each picture as the director asks for it.

▶ The control room of a television news program. Many of the items are broadcast live, so everyone has to think and act quickly. The newsreader in the studio receives instructions from the director through an earpiece which is not visible to the viewers.

director all the alternative pictures he or she can use. Only the central screen shows what is being transmitted.

The picture from each camera is shown on a separate monitor. One may show the program frontman, another the newsreader, while the third and fourth may show the empty demonstration area from different angles. A fifth camera may be all set up ready for the weather forecast.

Other monitors may be showing very strange things. A figure six on one comes from the film area to show that a film is loaded on the projector and is ready to go. This has to be started six seconds before the picture is ready, to reach the right speed, rather like an athlete running up to a jump.

Another monitor may show a clock. It is there as a symbol of the beginning of a video tape item. Like film, standard video needs a few seconds to get up to the right speed.

Another screen shows white lettering only. This comes from a minicomputer which produces words to identify people or things on the program. The words are superimposed on the pictures appearing on the screen. If slides are being used on the program, they are shown on the telejector machine.

**The Sound Engineers**

In a separate box near the control room are the sound engineers. Just as the director cuts from camera to camera, so the sound engineers control the microphones, add special sound effects and "fade" the sound in and out. It can be disastrous to leave a microphone on at the wrong time, as people will often say silly things if they think no one can hear them.

# Outside Broadcasts

It is easier to move television out of the studio than it looks. The whole gallery can be packed away in a vehicle called the scanner and driven to wherever the program needs to be. A big outside broadcast (OB) unit looks like a traveling circus.

**The Outside Broadcast Unit**
The staff are much the same as in a normal studio. They include engineers, the director, and lighting people, plus a very important character called the rigger driver. Rigger drivers drive the scanner, and the generator vehicle if there is no electric power at the site. They may also drive other back-up vehicles, some with video machines that can record material and edit it on the spot.

The engineers have an extra responsibility outdoors. They have to find a suitable spot for an antenna to send the pictures back to the television studio, from where they are sent out around the country. Some vehicles carry a dish antenna. If the signal is blocked by buildings, the engineers take the dish off and fix it in a high position.

**Sports and Instant News**
When television producers first discovered the excitement of taking the machinery outdoors, they went everywhere creating television "firsts" – broadcasting from mountains, aircraft and ships. Nowadays, OBs are usually kept for events which can't take place in the studio, such as sports.

One of the most interesting effects of OBs is that history sometimes happens live on television. Presidents die, men land on the Moon, and records are broken. The outside broadcast is a wonderful help to the news reporter, and this kind of outside broadcast is increasing. New developments in video electronics have brought smaller and better equipment so that OB units can get better pictures back to the studio much more quickly.

▼ A mobile ENG crew. Their recorded material can be rushed to the studios, or it can be edited on the spot and transmitted from the portable antenna.

▶ An outside broadcast unit. The scanner vehicle houses the monitors and a mixing desk. The antenna sends the pictures back to the studio.

**Electronic News Gathering**
The introduction of electronic news gathering (ENG) is replacing film coverage, particularly in North America. For ENG, a portable video camera is used. Sometimes it is used like film: material is shot, taken back to the studios and edited. At other times it is edited on the spot, and is very often transmitted live into news bulletins.

ENG trucks carry a pop-up antenna to send the picture back, and editing machines for the video tape. Two cameras can be brought out of the truck on cables or they can wander free using battery power. This is called the ENG mode.

ENG is very useful during a difficult situation such as a riot. While the truck is parked safely in a side street, the cameraman and his sound recordist dash around taking video pictures. It takes only a few seconds to take the video tape out of the recorder, or "deck", and as soon as the crew get back to the truck they beam the pictures back to the television station.

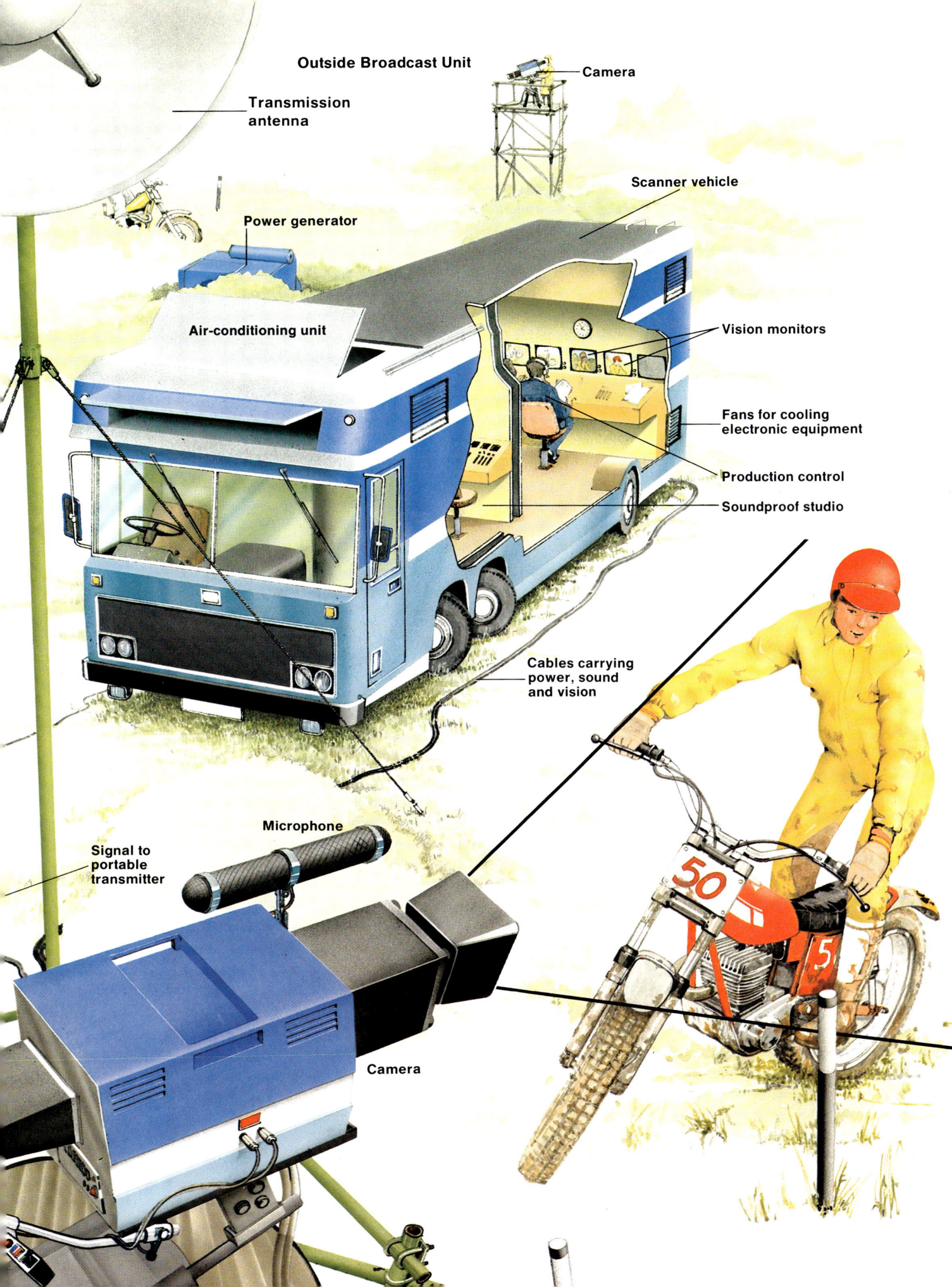

# Film and Video Tape

There is a well-known television story about the director whose program was going wrong. Pre-recorded film and video items did not appear. He kept completely calm in the chaos and finally leaned forward and said to everyone listening: "Okay, anyone who's got anything reeled up, run it!" He got the closing credits only four minutes after the program had begun!

## Pre-recorded and Live Programs

Television is almost too good to be true, with perfect presentation and actors who don't forget their lines. It is, of course, so good because most of it is recorded, not "live", and the mistakes are edited out long before the programs are broadcast. Even live programs such as newscasts have many pre-recorded items.

## Recording on Film

Film is the oldest recording medium used in television. The normal film crew includes a cameraman and assistant, a sound man (with an assistant if needed) and one or more people for lighting. They usually use 16-millimeter-wide film (twice as wide as home movie film) of two different kinds.

Ektachrome is the cheaper and easier film because only one developing process is needed, and it produces an image which can be broadcast directly. There is only one original

▶ A video editor never cuts into the tape. He or she re-records the wanted sections on to a new tape by using two machines called an edit pair. Making a series of smooth edits is a job for professional machinery. With home video equipment it is very difficult to make edits that don't make a noise or "jump".

which is cut and transmitted. The disadvantage of Ektachrome is that the image can easily become scratched and dirty with use.

The other 16-millimeter film is called Eastman Color. This produces a negative image from which a "cutting" print is made. When the print has been altered and approved, the editor goes back to the original and gets a brand-new print. He or she matches it with the cut version by matching numbers on the edge of the film. Look closely when you are watching television and you may be able to spot the difference between Ektachrome used in news programs and Eastman Color used for longer films with beautiful scenery.

## Storing Pictures on Tape

A more recent way of storing television pictures is electronically, on video tape. There are different sizes of tape which produce pictures of varying quality. Many engineers will say that the wider the video tape, the better the picture.

At the moment, television companies use a mixture of machines, some with two-inch (50-millimeter) tape, some with one-inch (25-millimeter) tape, and some with three-quarter-inch (20-millimeter) tape. Sound and pictures are on the same tape and the editing process is different from film.

## Editing Film and Video Tape

The film editor works by using a machine that cuts the film into strips and sticks it back together again. Unwanted film is simply cut out. Video editing is done by re-recording sections in the right order on to another machine. The tape itself is never cut. When you have finished with the material on tape, it can be erased and the tape is ready to be used again.

Film, of course, is permanent and is therefore much more expensive than video tape. On the other hand, film cameras and editing machines are much tougher and last very much longer than video equipment. And there's the argument about the quality of the picture. Although the video picture is clearer, directors still prefer to use film to achieve soft, romantic effects.

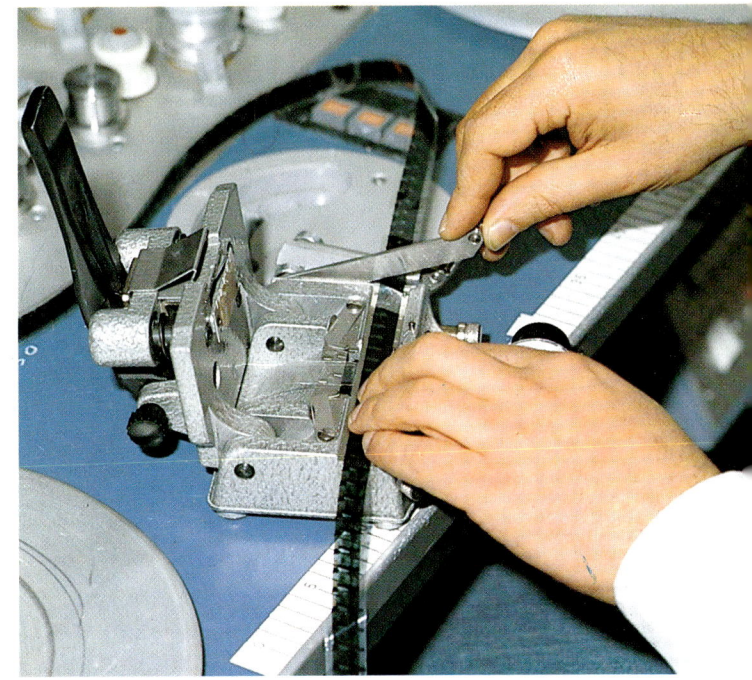

▼ A film editor cuts the film and places the edges together in a joiner. The film is held in place by sprocket holes in the joiner which match the holes in the film. The editor puts special sticky tape over the edges of the film and punches the joiner down to stick the film together.

▲ The film editor also matches the pictures with the soundtrack. To help with this, a clapperboard picture appears at the beginning of each shot. As the board closes, the editor finds the matching noise on the soundtrack. All the pictures and sounds which follow will also match.

# Satellites and Cable

The family of the future will have many more television channels than we have at present. There are comparatively few because national television channels are duplicated all around the country and this uses up the available frequencies. To have more channels, other methods must be used, such as signals transmitted by satellite or channelled along cables into the home.

## Beaming Pictures by Satellite

Television people call satellites "birds". They use them all over the world, especially for news and sport. Satellite transmission is like passing a ball to your partner by bouncing it off the ceiling. The television station sends its picture to a ground station, which sends it up to a satellite. This bounces it back down to the ground station in whichever country the signal is going to. From here it passes along ground lines or it can even be bounced along more dishes (called microwave dishes) to the television station. Then it is sent out to the transmitters and the viewers.

International satellites are like aircraft with bookable flights. Most of the time available on them is booked by the big three US television networks: NBC, CBS and ABC. Quite often they book more time than they need, to try to stop their competitors. Smaller companies in the rest of the world can "borrow" time. Suppose an American network has a booking to get pictures from a sporting event in Turkey. European countries can get a "downleg" of these pictures to their own ground stations as they pass over on the way to the United States.

Direct broadcast satellites (DBS) work in a similar way, but produce a much stronger signal. To receive pictures from an international satellite you need a receiving dish bigger than a house. To get DBS pictures, a dish about four feet (one meter) across is big enough. You need permission to place it correctly in the yard to work efficiently, as well as an expensive alteration to your television set.

## Cable Television

A big extension of cable television is also planned. Cable isn't a new idea. It is used throughout the United States. This kind of distribution is particularly useful in mountainous areas because the mountains interfere with airborne signals.

Copper coaxial cables carry television pictures and sound as electrical signals. These signals have to be boosted every few miles. Now there is a new kind of cable called optical-fiber cable. This is made of bundles of stretched-out glass which can carry signals uninterrupted for longer distances. The electrical signals are converted into pulses of light which travel along the thin glass fibers inside the cable. Each cable can carry many television channels.

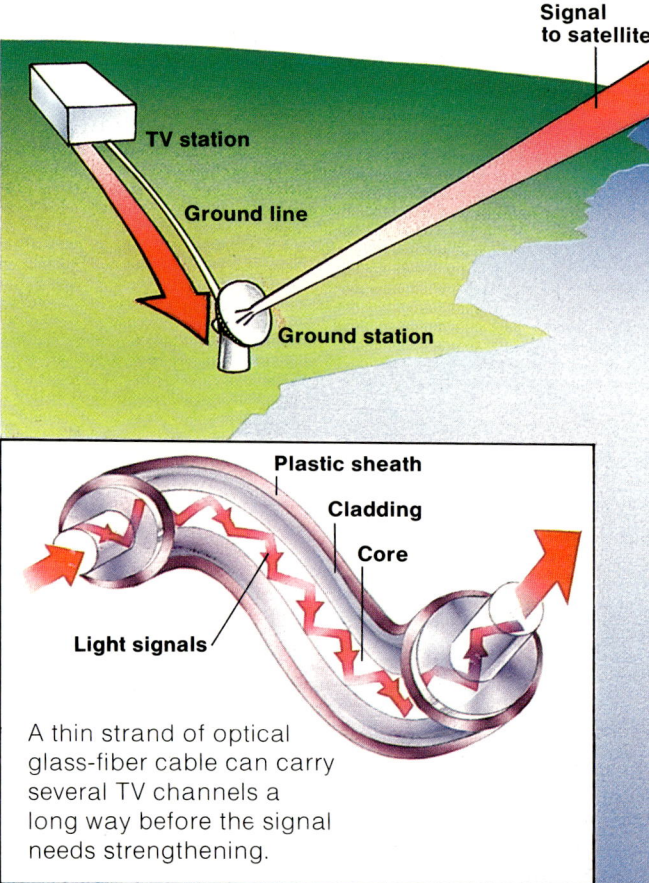

A thin strand of optical glass-fiber cable can carry several TV channels a long way before the signal needs strengthening.

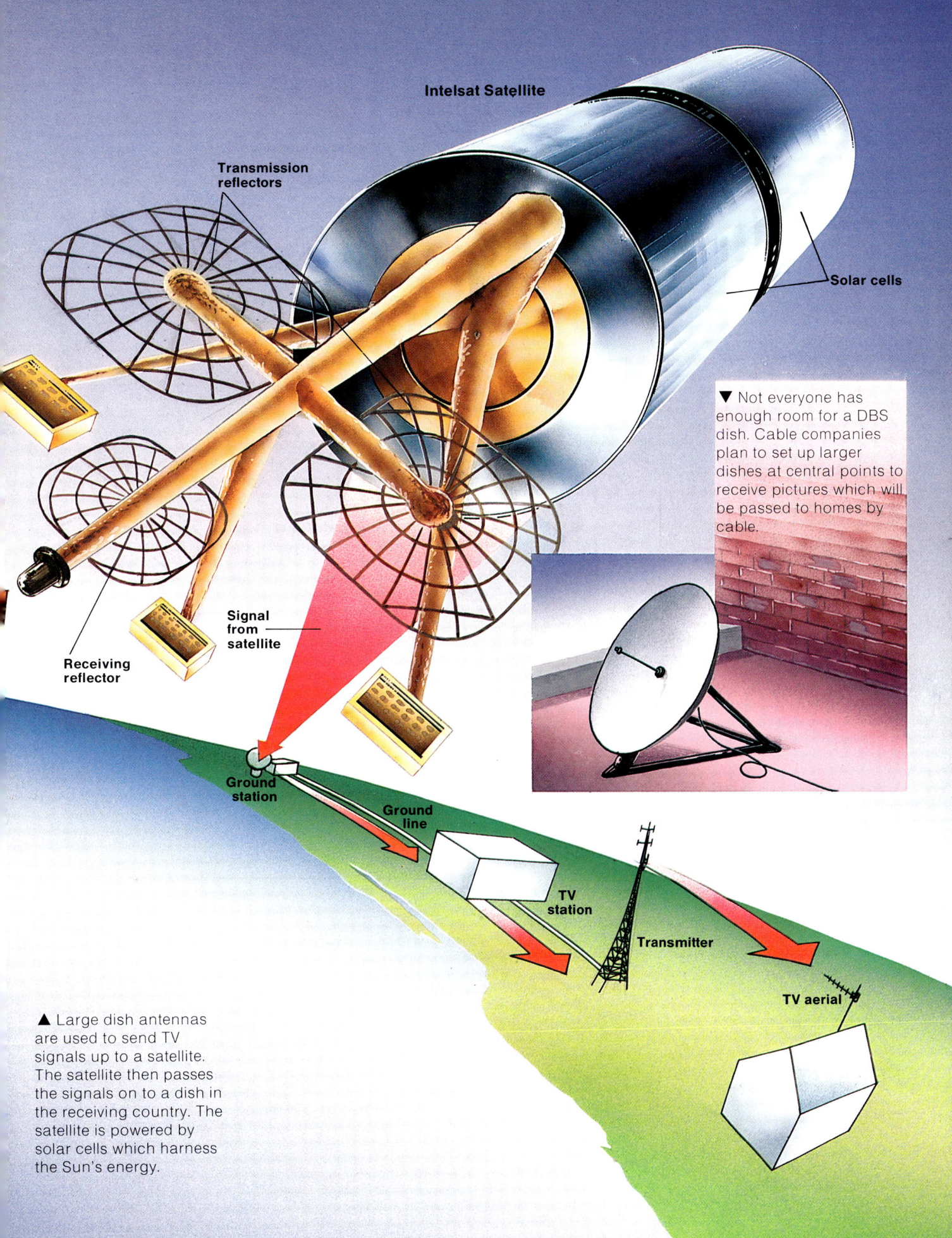

# Developments in TV

A television set is packed with complicated electronic equipment to decode the signals and show us a picture we understand. It is difficult to make smaller and thinner sets because of the size and shape of the cathode ray tube. To make flat screens, the tube has to be redesigned, or a different method has to be invented to decode the picture signal.

## Making Screens with Lasers

A lot of time and money have been spent looking for a replacement for the cathode ray tube. A French team has made great progress by using a coating of silicon on glass. (Silicon is the material used for making microchips.) They use a ruby laser to turn areas of the silicon coating into crystals. Finally, using normal microchip methods, they turn the crystals into transistors, which are tiny electrical switches.

## Wristwatch Television

Wristwatch television is also on its way. People will be able to buy miniature sets that show them breakfast television on the way to work. These sets will also tell you the time and date, and you may be able to play space invaders with them too. This may be a short-lived craze. For example, you can print books so small that you need a magnifying glass to read them. Tiny television could be just as frustrating.

▲ The eidophor system uses a strong light beam and slatted mirrors to project television pictures off a thin film of oil on to a large screen. Because of the high cost, this system is not practical for use at home but is used in places such as sports stadiums.

▶ In present 3-D TV, two identical images are seen together through special viewing spectacles (1). Another method will use a double television tube to project two images (one for each eye) on to a grooved lens (2). The grooves direct the right information to each eye.

**3-D Television**

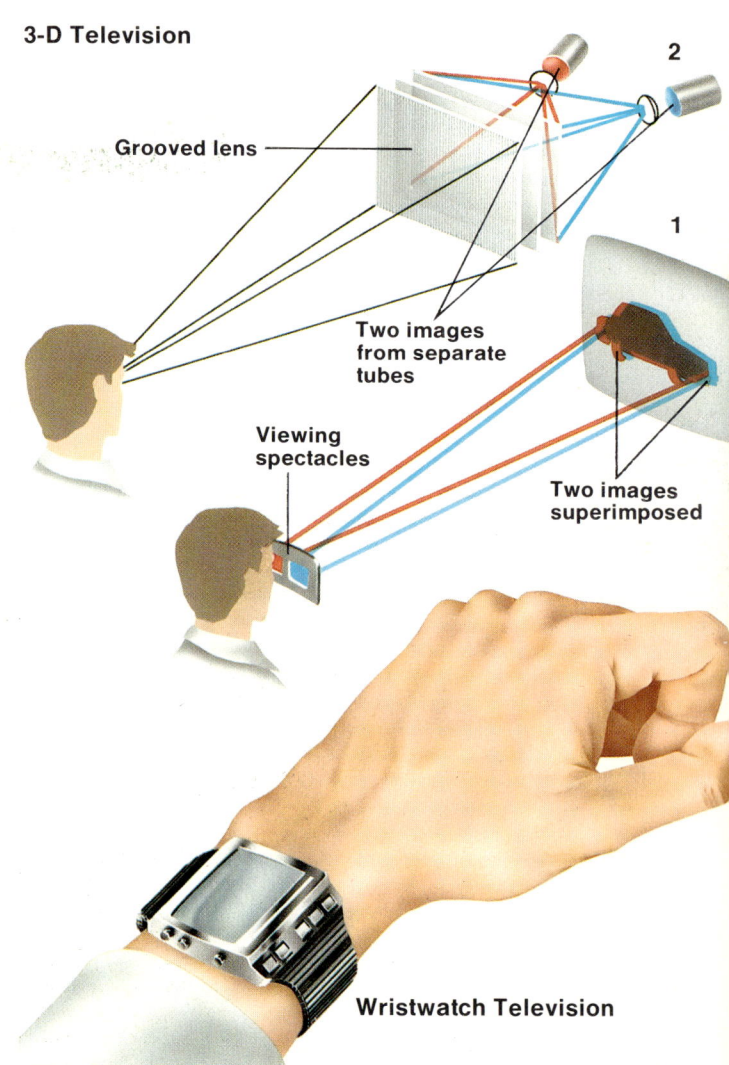

Wristwatch Television

## 3-D Television

Three-dimensional television works by pointing two cameras at the same object from different angles, then superimposing the two pictures. You can see depth, and objects seem to come out of the screen at you.

Three-dimensional photography was tried out many years ago in the cinema and has recently become popular again, now that new equipment has been developed. It is particularly in demand for horror films, when it is really exciting to imagine a vampire coming out of the screen towards you.

The major problem with 3-D photography is that the eyes need help to put the two separate images together, and at the moment viewers have to wear special glasses. The problem is worse for television than for the cinema with its big screen. With a small screen, you have to keep your head still or the picture breaks up.

The latest 3-D experiments are designed to get rid of the glasses that viewers have to wear. Instead, engineers are trying to use a double television tube with a special grooved lens in front to create an instant picture from the two.

## The Effects of Television

The biggest future development is not really technical, but about how our lives may change because of television. Some doctors are already worried about children who spend too much time playing with home computers and video games all by themselves. If people can do a lot of work at home using television, and have no need to go out unless they want to, will they perhaps become people who meet only through television?

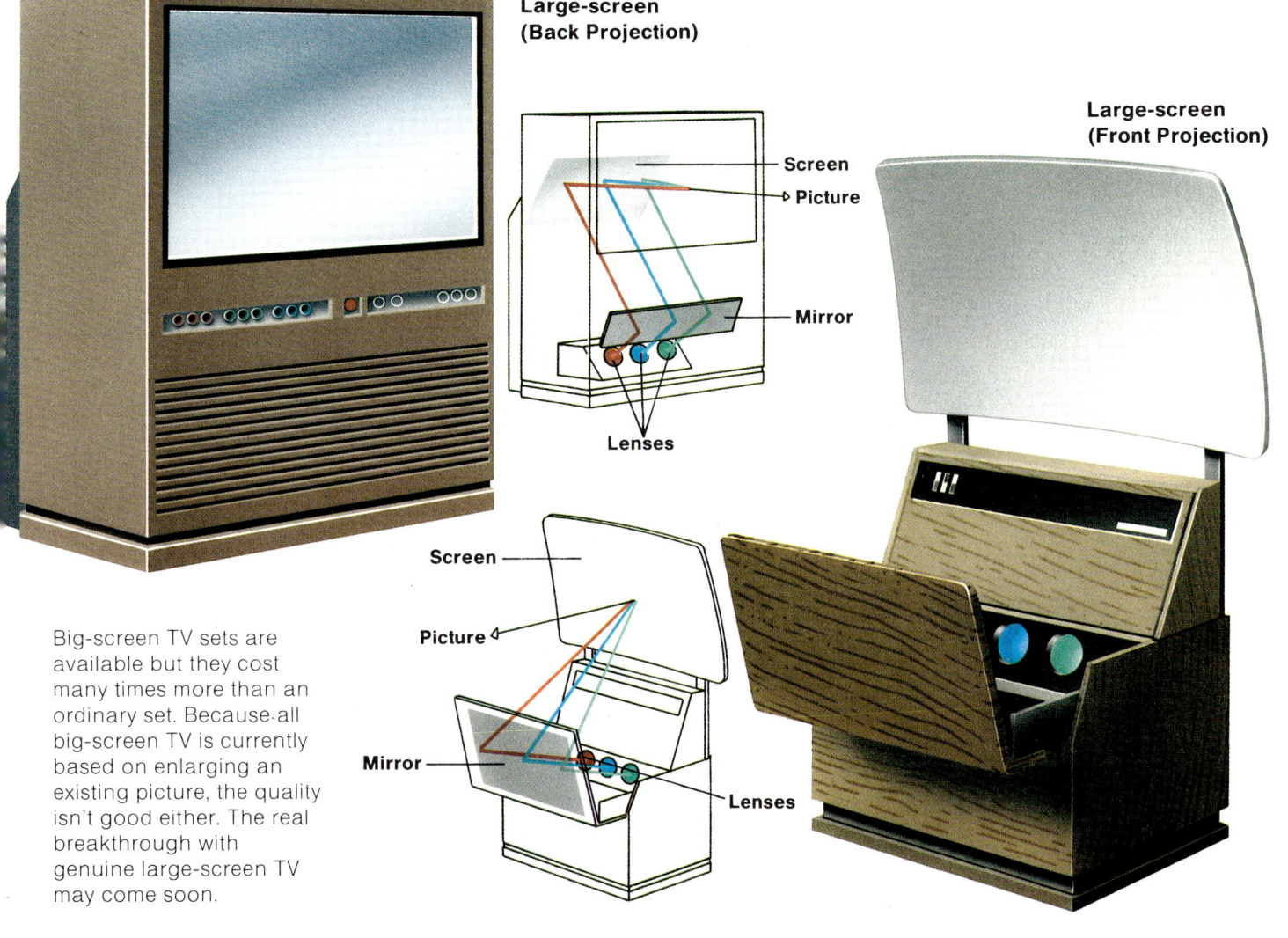

Big-screen TV sets are available but they cost many times more than an ordinary set. Because all big-screen TV is currently based on enlarging an existing picture, the quality isn't good either. The real breakthrough with genuine large-screen TV may come soon.

# Video Recorders and Video Disks

The first home video machines appeared in the early 1970s. They used one-hour tapes and had simple controls. Now video machines can be programmed up to 14 days in advance, and can record on different channels.

### Recording on Tape
There are several types of home video-tape systems. Unlike records or sound cassettes, however, you cannot switch video tapes from one system to another, even though the machines work the same way.

Picture recording is similar to sound recording. They both use magnetic tape. A recording head in the video recorder transfers an electronic signal to the tape by altering or disturbing the magnetic field on the tape. The signal is replayed by doing the same thing again, but in reverse.

As we have seen, a picture signal is more complicated than a sound signal and it needs more space. So video tape is wider than music tape. The head records the signal in a diagonal pattern across the tape, in strips close together. This enables you to get more on. You could record the picture in a straight line, but it would take 20 miles (33 kilometers) of tape to make a one-hour recording.

### Video Disks
The video disk is different from video tape in several ways, the most important one being that you can't record on a disk at home. It is rather like the difference between records and cassette tapes for music. Disks have been used professionally for several years for sports "action replays", but are new to most people.

The leading system of picture recording on disk uses lasers. The disk is made of plastic and is printed with tiny pits which hold the picture and sound information. Instead of a needle touching the disk, a laser beam is used to read the information in the pits. The beam is constantly directed towards the pits by a series of mirrors and prisms. Most people agree that the picture quality of disks is better than that of video tape.

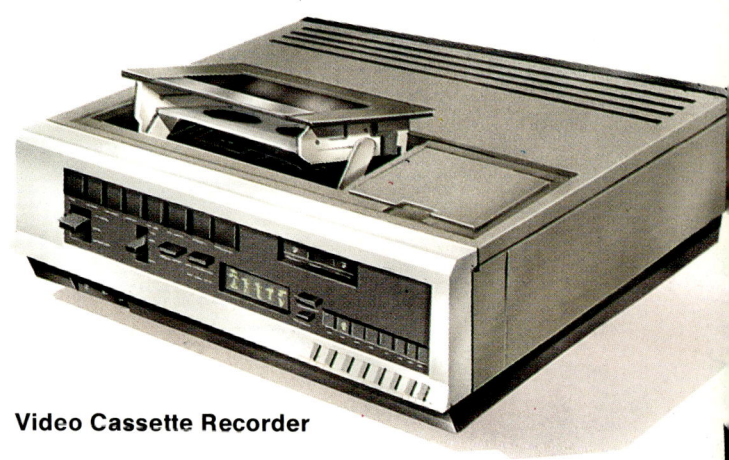
**Video Cassette Recorder**

### Making Disks and Tapes
Disks should be quicker and cheaper to make than pre-recorded tapes because each disk is stamped out with a complete program on it, rather like an LP record. Video tapes are produced by copying another video tape. This process takes as long as the program lasts, and cannot be speeded up. Sometimes, because of careful checking at each stage of production, disks may also take a long time to make too, so it is difficult to say which system has the most advantages.

**Video-disk Player**

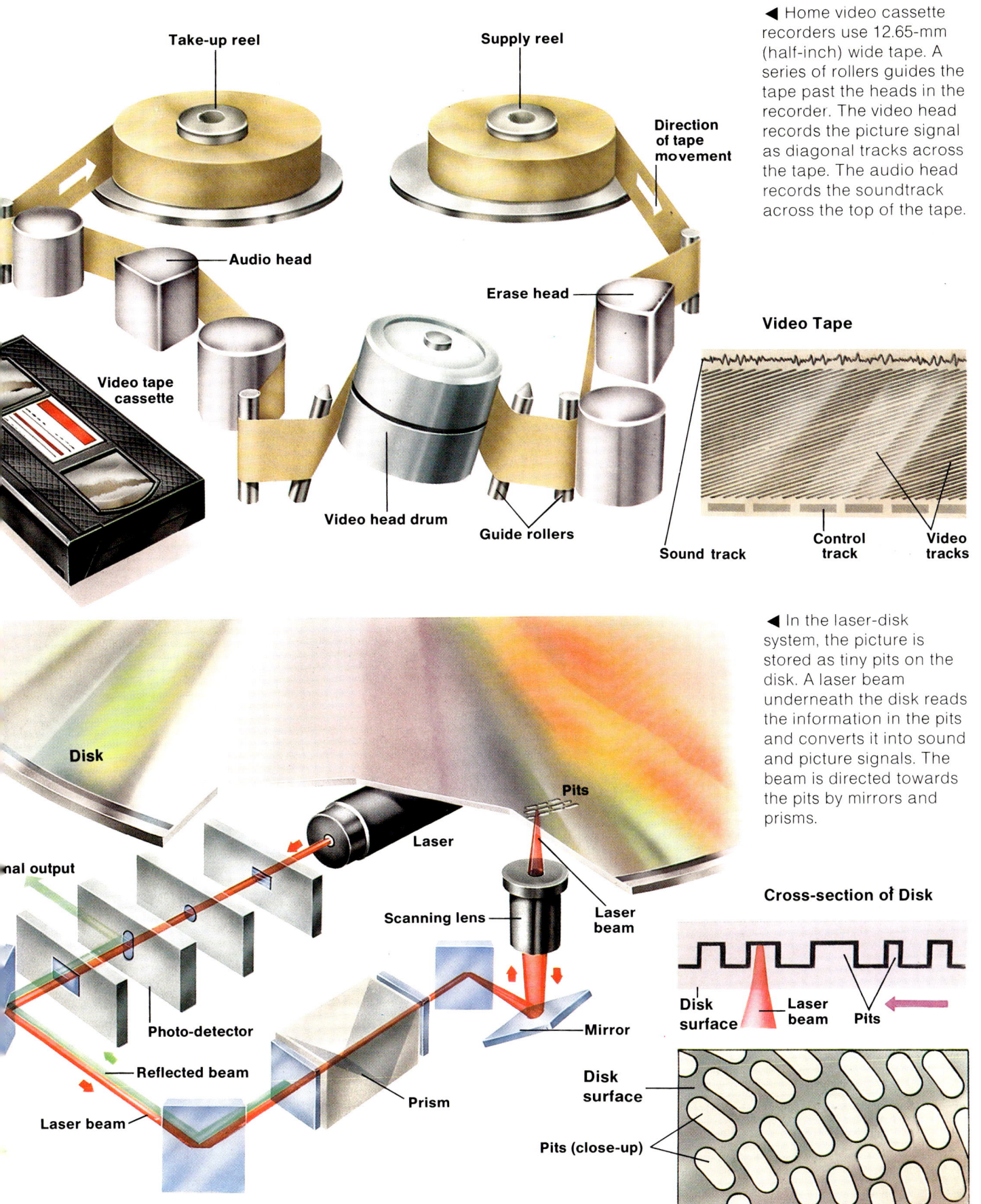

◀ Home video cassette recorders use 12.65-mm (half-inch) wide tape. A series of rollers guides the tape past the heads in the recorder. The video head records the picture signal as diagonal tracks across the tape. The audio head records the soundtrack across the top of the tape.

◀ In the laser-disk system, the picture is stored as tiny pits on the disk. A laser beam underneath the disk reads the information in the pits and converts it into sound and picture signals. The beam is directed towards the pits by mirrors and prisms.

# Making a Video: Equipment

Most home video recording machines can be fitted with a camera that can be plugged into it through a camera adapter box. This system isn't portable, so you may have to stay in the house or use your camera in the yard with the help of an extension lead. Portable cameras and recording decks to match are available, but they cost more than home movie cameras.

### The Advantage of Video
There is a great advantage to using video equipment instead of film. If you make mistakes with film, you don't find out what's wrong until it has been processed, and that may mean you have wasted a lot of time and money. With video, you can experiment with shots to see how they look as they are being taken. You check the shots by replaying them on the recorder. If there's anything wrong, you can erase the shots on the tape and start again. The same tape can be used over and over again until it's right.

### The Camera
There are two important things to remember. Read your camera instructions carefully and, if you can't understand any of them, get expert advice. Never take things to pieces to "make them work". Second, a video camera is not very strong compared with a still camera or home movie camera. It has a delicate tube like those in television studio cameras, and you can break the tube by dropping or bumping the camera. You can "burn" the tube too. Burning happens when the iris, or aperture, is left open and the camera is accidentally pointed at the sun or at a strong light.

### Microphones
Sound varies from camera to camera. If possible, always plug in a separate microphone if your system allows it. The "mike" built into the camera is a very poor method of collecting sounds.

### Advance Planning
Making a video is not really something you can do by yourself. You need several friends to help you. Give everybody a job before you start. You will need actors, a camera operator to shoot the pictures, and a producer to write the script and see that everybody is ready. The sound recordist's job is to hold the microphone in the right position. The properties and continuity person checks all the details. The engineer watches the television screen and makes sure that the equipment is working perfectly.

Work to a definite plan because you can only pause between shots once you start the camera. The picture will blur every time you switch off and on again.

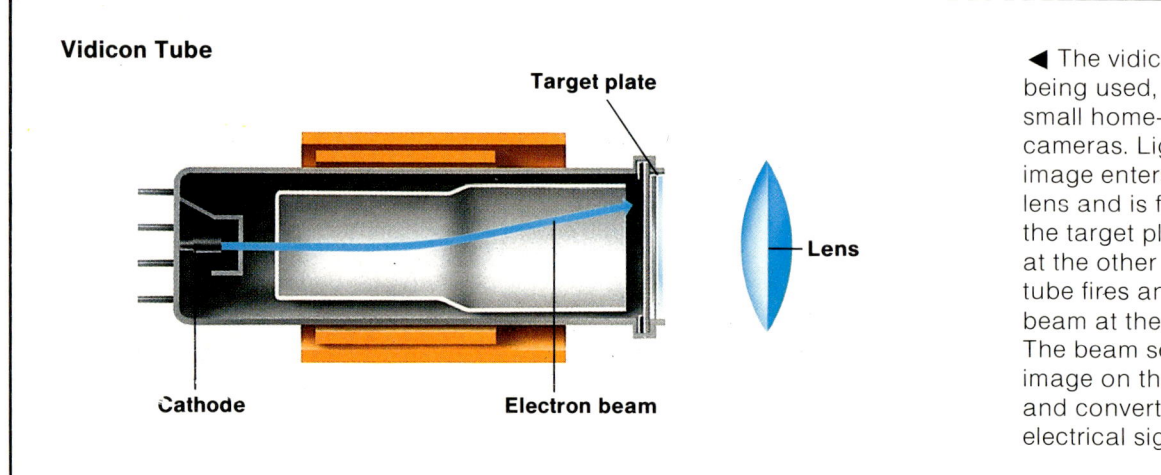

◀ The vidicon tube is still being used, particularly in small home-video cameras. Light from the image enters through the lens and is focused on to the target plate. A cathode at the other end of the tube fires an electron beam at the target plate. The beam scans the image on the target plate and converts it into an electrical signal.

Video cameras are simple to operate. There are three main controls to learn about. The iris, or aperture, controls the amount of light going into the camera. This control may be automatic or manual. The focus control is used to make sure that each shot is sharp. It is usually manual. The zoom is a way of closing in to or pulling out from the subject without losing the sharpness, or "focus". The zoom control may be manual or automatic.

Another important control is the "white balance". This is used to adapt the camera to different lighting conditions. On cheaper cameras, this control will be manual. Always check each shot on your TV screen and adjust the control according to the maker's instructions.

# Making a Video: Shooting

The beginner should start making videos outside in the yard on a sunny day. Connect the camera to the power points inside the house with an extension lead. Always ask an adult to check that the plugs and connections are safe.

## Using the Microphone

The sound recordist's microphone should be connected to the video recorder inside the house via an extension lead. The best way to shoot drama is to tie the microphone to a stick. The sound recordist can then hold it over the heads of the actors, taking care that the mike is low enough to record the sound but doesn't get into the picture. The continuity person should check that the actors know their lines.

## Using the Camera

The camera operator should try to sit opposite the actors, or lean against a wall to steady the camera. The less jerky the camera movement is, the more professional-looking your video will be. Make sure the sun isn't behind your subjects or the picture will come out looking black.

The camera operator should also practice different shots such as holding, panning, zooming and tilting. For the hold, steadily point the camera at the subject without moving the camera or the zoom control. To pan, move the camera horizontally across the subject from left to right or right to left. The tilt is similar to the pan, but is a vertical movement. To zoom, you usually press the zoom button on the camera. This closes in to a subject, or pulls out from it. Zooming should be used only when necessary. Even among professionals there is a tendency to zoom too much.

Because the picture wobbles for the first few seconds after the camera has been switched on, you should shoot your sequences in the correct order. Use the pause button to alter a shot when you cannot pan or zoom to change shot.

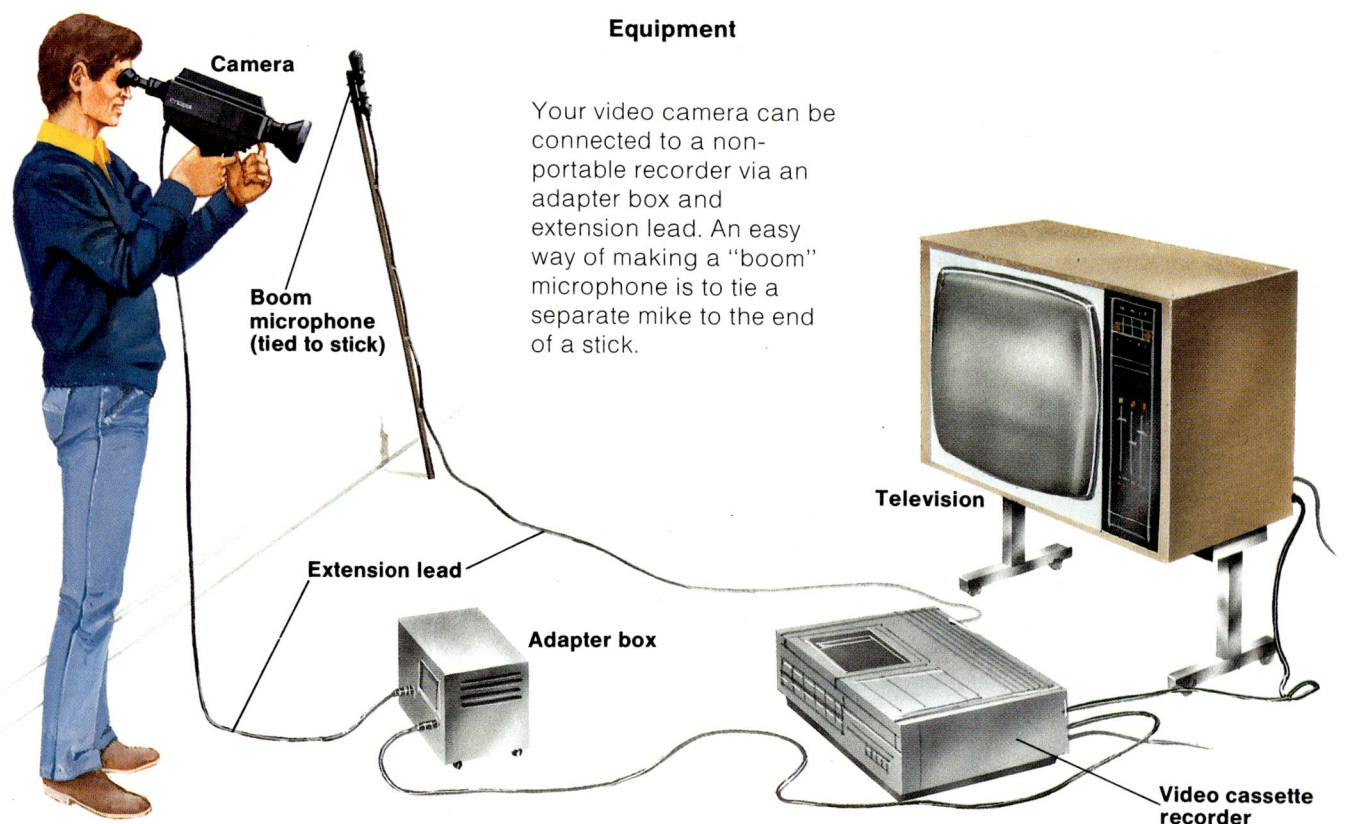

**Equipment**

Your video camera can be connected to a non-portable recorder via an adapter box and extension lead. An easy way of making a "boom" microphone is to tie a separate mike to the end of a stick.

# Camera Shots

### Long shot

▲ To hold, point the camera at the subject steadily. A long shot shows background detail.

### Mid-shot

▲ A mid-shot shows more of the subject and cuts out some of the background.

### Close-up

▲ A close-up cuts out the background and concentrates on the subject.

### Panning

(1)

(2)
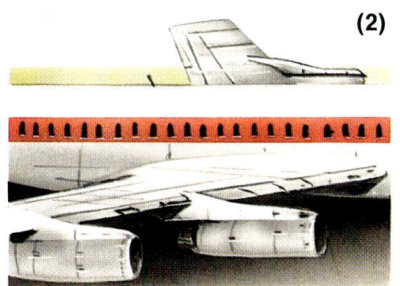

(3)

### Tilting

(3)

(2)

(1)
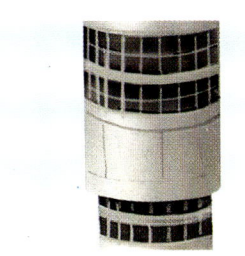

▲ To pan, steadily move the camera horizontally across the subject. Do not make the panning movement too quickly.

▶ The zoom enables you to close in to or pull out from the subject without moving the camera. The zoom should be used only when necessary.

◀ The tilt is similar to the pan, but is a vertical movement up or down, and is good for tall subjects.

### Zooming

# Video at Work

The most common use for video equipment at work today is security. We have become quite used to being watched by cameras in stores. The old-fashioned store, where a sales assistant stood behind a counter and handed the customer each item, didn't need a camera "spy". But modern supermarkets are designed to tempt us into picking things up and paying for them later. Unfortunately, this means that some people are tempted into picking things up and not paying for them at all!

## Security and the Law

Security cameras are small, simple and usually give black and white pictures. They are connected to a monitor, not a television receiver, which can be watched by the shop security staff. Camera systems are also widely used in banks.

Video technology has now been introduced into the courts, especially in North America. Cameras are good "detectives", and the evidence they provide shows the criminal

▶ The police now use video to help them with traffic control. They can see from a glance at the screens how traffic is moving, and can take immediate action if necessary.

▼ By using closed-circuit television, this worker at a nuclear power station can keep a constant check on the storage of dangerous radioactive materials.

actually at work on the crime.

Video cameras can also protect suspects who have been arrested. An interview at the police station can be video-taped with a clock in the background to prevent the tapes being edited or altered in any way.

Domestic security also can be improved with video cameras. Some local councils have installed them at the entrances of apartment blocks to cut down vandalism and attacks on old people. These cameras are connected to a separate cable channel on television sets in the block, so that the residents can see who's coming and going.

## Teaching and Training

Teaching by video will become as important as security. Operations, for example, can be watched on a screen by many medical students at once. Doctors would also find video records a great help instead of having only written notes about their patients. Sick people in hospital, and babies in incubators, can be watched on a screen when the nurse isn't actually in the ward.

Business companies and other organizations can record their work and use the tapes to train new staff. Sports trainers and dance teachers already find video very useful. Movements can

be recorded in close-up detail and then played back slowly to see how the performance can be improved. Sports teams can use video to re-live a game and look at all the mistakes they made. If a complaint is made after a horse race, video can be used to see whether a jockey has behaved properly.

Estate agents have started using video tapes to sell property. Instead of looking at pictures and then going to see a lot of houses, people can view a video tape to give them a better idea of what the houses would be like to live in. They can "visit" many houses in half an hour on video before choosing the few they are serious about visiting.

## The Future

Perhaps salesmen of the future who come to the door will invite you to see a tape, instead of carrying heavy samples of the goods. Catalogs may be transformed from heavy volumes into television programs, either replayed by special services on television or given to you to look at in the store.

With all these possibilities, could we become a society where everybody is "watched", whether they like it or not? It is a worrying question to which there is no simple answer. We will have to make sure that the reasons for watching us are very good ones.

◀ Dancers in training or rehearsing new roles can be recorded on video. By playing back the video later, they can study their performance and improve on it.

▼ Many organizations now use video for training new recruits. By recording lectures and demonstrations on tape, they can be replayed many times over to different groups.

# Glossary

**Barn doors** Hinged black flaps attached to studio lights. They can be moved forward and backward to increase or decrease the amount of light.

**Bird** Television slang for "satellite".

**Booster station** Television signals grow weaker as they travel away from transmitters. Booster stations pick them up, strengthen them and pass them on.

**Cable TV** A method of carrying pictures and sound to television sets by cable. The cable may be buried underground or strung on telegraph poles.

**Carrier waves** These are generated by TV engineers to carry sound and picture signals through the air.

**Cathode ray tube** A glass vacuum tube found in television sets. It has a broad end (the screen) and a narrow end containing electron guns.

**Channel** Television signals travel on a limited number of frequencies through the air. Each TV channel is given a different frequency.

**Channel programmer** An electronic control unit on a video recorder. It enables the video recorder to stop, start and change channels on a time control.

**Chrominance signal** The color signal produced by a TV camera. It is made up from the primary colors of light: red, green and blue.

**Closed-circuit TV** A system for viewing without transmission being involved. A TV monitor, rather than a TV receiver, is used to display the pictures.

**Coaxial cable** A cable that consists of a tube of electrically conducting material, inside which is a central conductor. The central conductor is insulated from the tube. These cables are used to carry very high frequency signals such as those used in televison.

**Color balance** A color mechanism on a TV camera which enables it to be adjusted for lighting differences indoors and outdoors.

**DBS** Short for direct broadcast satellite. A DBS sends out a signal strong enough to be received by a domestic dish antenna.

**Earpiece** A listening device worn by the presenter of a TV program. It is not visible to the audience but enables the director in the gallery to talk directly to the presenter.

**Eastman Color** 16-millimeter-wide film which produces a negative image from which prints are made for transmission.

**Ektachrome** The common name for 16-millimeter-wide film which produces one positive image only.

**Electron guns** In a television set, electron guns shoot electrons through the slots in the shadow mask on to the phosphor dots on the screen. This makes the phosphor dots glow with color to produce a picture. In a television camera, the electron guns fire electrons at the target plate to scan the pattern of the image and turn it into electrical signals.

**ENG** Short for electronic news gathering. The news team carry a portable electronic camera linked by cable to a video recorder.

**Flats** Movable pieces of scenery used in a TV studio.

**Floor manager** The human link between the

TV gallery and the studio floor. The manager uses headphones to listen to instructions from the director in the gallery.

**Gallery** The control room at a TV studio which contains the mixing desk and monitors.

**Luminance signal** The black-and-white signal produced by a TV camera. Sometimes this is produced by a separate tube in the camera, sometimes by the three color tubes.

**Microphone** Used for transforming sounds into electrical signals. In this form, sound signals can be recorded and then played back through a loudspeaker.

**Millimeter** The traditional unit of measurement for film width. Film used for recording TV programs is usually 16 millimeters wide.

**Mixing desk** A control center into which different picture sources can be fed for displaying on monitors. The director selects each picture for transmission and changes it when he or she likes.

**Monitor** A screen used for closed-circuit picture display. Monitors, unlike TV sets used in the home, do not receive broadcast pictures via an antenna.

**Optical-fiber cable** Thin glass fibers bunched together to make a cable capable of carrying a very large number of TV channels. TV signals travel along the cable as pulses of light. They can travel long distances before they need boosting.

**Outside broadcasts** TV programs made outside a television studio. Outside broadcast programs can be fed into the network by using a portable transmitting antenna.

**Phosphor dots** Used to coat the inside of a TV screen. The phosphor dots glow with color to form a picture when they are hit by a stream of electrons.

**Presenter** A person who, in many television programs, appears on the screen to introduce various items and guests. He or she is often called the frontman, or linkman.

**Production assistant** The person responsible for program timings in the gallery. He or she makes sure that everything starts and stops on time and that pre-recorded items mix smoothly with the "live" ones.

**Rigger driver** The person who drives the outside broadcast vehicle (the scanner) and does many of the hard jobs involved in setting up an outside broadcast.

**Three-dimensional TV** A way of making an image more life-like by showing it from two different directions and superimposing the two pictures.

**Tracking** Moving a TV camera forward, backward or sideways without using the zoom.

**Transmitter** An engineering center with a tall mast which is used to send out television signals through the air.

**Tuner** The mechanism inside a TV set which is used to select different signals and feed them to the channel buttons on the set.

**Video cassette** An enclosed reel of magnetic tape for recording on commercial or home video recorders. The tape carries pictures and sound.

**Video disk** A flat disk on which pictures and sound are recorded. It is replayed on a video-disk player.

**Vision mixer** The person who sits at the mixing desk in the gallery next to the director. The vision mixer operates the switches controlling the picture sources.

**Zooming** Closing in to, or pulling out from, a subject without moving the TV camera. This is done with the aid of a zoom lens.

# Index

Note: Page numbers in *italics* refer to illustrations.

**A**
Antenna 18, *19*, 22, 23
Aperture 28, *29*

**B**
Baird, John Logie 8
Barn door 14, 34
Big-screen TV 25
Birds 22
Black and white television 11, 13
Boom microphone 14, *30*
Booster station 13, 34

**C**
Cable television 22, *23*, 34
Camera: security 32, *32*; television 12, *12–13*, 18, *19*; video 18, 28, *28, 29*, 30, *30, 31*
Cameraman 13, 14, 20
Carrier waves 13, *13*, 34
Cassette 26, *27*, 30
Cathode ray tube 10, *10*, 24, 28, 34
Channel 22, 34
Channel programmer 34
Chrominance signal 13, 34
Clapperboard picture *21*
Closed-circuit camera 8, 9
Closed-circuit screen 16, *32*
Coaxial cable 34
Color television 10, *11*, 12
Continuity person 28, 30
Control room 14, 16, *16, 17*, 18
Court evidence on video 32
Cutting print 21

**D**
Dancing recording 32
DBS *see* Direct broadcast satellite

Dichroic mirror *12*, 13
Director 14, 16, *16*, 17
Direct broadcast satellite 22, 23
Dish 18, 22, 23
Disk 8, *9*, 26, *27*

**E**
Earpiece 34
Eastman Color film 21, 34
Ectachrome 34
Editing 20, 21, *21*
Edit pair *20*
Eidophor system 24
Ektachrome film 20
Electron 12, *12*
Electron gun 10, *10*, 11, 12, *12*, *28*, 34
Electronic news gathering 18, *18*, 34
Electronic signals 10, 12, *12*, 22, *22*
ENG *see* Electronic news gathering
Engineer 16, 17, 18, 28

**F**
Family of the future 8–9, *8–9*, 22, 33
Film 20, 21, 28
Film recording 20
Flats 14, 34
Floor manager 14, 34
Future television and video 8, *8–9*, 25, 33

**G**
Gallery 14, 16, *16, 17*, 18, 34
Ground station 22, *22*

**H**
Headphones 8, *9*, 13, 14, *14*
Holding 30
Home security 32

**I**
Intelsat satellite 23
International satellite 22, 23
Iris 28, *29*

**J**
Joiner *21*
Joystick control *13*

**L**
Laser beam 26, *27*
Lecture recording 32
Lens 12, *12*, 14, 25
Light *11*, 12, *12*
Lighting 14, 20
Lighting panel 14
Live program 18, 20
Loudspeaker 10
Luminance signal 13, 34

**M**
Mask 10
Medical use of video 33
Microphone 14, 17, *19*, 28, *29*, 30, 35
Microwave dish 22
Millimeter 35
Minicomputer 17
Mixing desk 16, 35
Monitor *13*, 14, 16, *19*, 32, 35

**N**
Networks 22
News program 14, 17, 18, 20, 21
Nuclear power station 32

**O**
Optical-fiber cable 22, *22*, 35
Outside broadcast 18, *19*, 35

**P**
Panning 30, *31*

Phosphor dots 10, *10*, 11, *11*, 35
Picture 10, *10*, 21
Picture lines 10
Picture recording 26
Picture signal 10, *10*, 13, *13*, 26
Pits 26, *27*
Police video 32, *32*
Power generator *19*
Pre-recorded program 20, 26
Presenter 16, 35
Production assistant 16, 35

**R**
Recording: television 20; video 26, 28
Relay station *13*
Rigger driver 18, 35
Ruby laser 24

**S**
Satellite 35
Satellite dish receiver 8, 9, 22
Satellite transmission 22, *23*

Scanner 12, *12*, 18, *19*, 27
School of the future 8
Screen 8, 10, *10*, *11*, 24, 25, *25*, 32
Security camera 32, *32*
Signal transmitter 13
Silicon 24
Solar cell *23*
Sound engineer 17
Sound recordist 14, 28, 30
Sound signal 13, *13*
Soundtrack *21*, *27*
Spectacles *24*, 25
Sporting events 22, *24*, 26, 33
Staff training 33, *33*
Storing pictures 21
Studio 12–13, *12–13*
Supermarket 8, 32

**T**
Tape 17, 18, *20*, 21, 26, *27*, 28
Target plate 12, *12*
Teaching: by television 8; by video 33, *33*
Telephone communications 8
Television 10–11, *10–11*, 30

Three-dimensional TV 25, 35
Tilting 30, *31*
Tracking *13*, 14, 35
Traffic control 32
Transmitter 10, 13, *13*, *19*, 22, 35
Tube 12, *12*, 13, 25, 28
Tuner 10, 35

**V**
Video 18, 26, *26*, *27*, 35
Video cassette 35
Video disk 35
Video-making 28, 30, *30*, *31*
Video mixer 35
Vidicon tube *28*
Viewing spectacles *24*, 25
Vision mixer 16

**W**
White balance 29
White light 11, *11*
Wristwatch television 24, *24*

**Z**
Zooming *13*, 14, *29*, 30, *31*, 35

---

## ACKNOWLEDGEMENTS

Front cover: bottom left, SFP/Jean-Claude Pierdet; bottom right, Gretag Elektronik GMBH; contents page: TVS; 14–15: SFP/Jean-Claude Pierdet; 16: TVS; 17: ITN; 18: SFP/Jean-Claude Pierdet; 20–21: TVS; 24 left: Gretag Elektronik GMBH; top: Metropolitan Police, New Scotland Yard; 32 bottom: UK Atomic Energy Authority; 33 top: ZEFA; 33 bottom: Sony (UK) Ltd.

Picture research: Jackie Cookson

The publishers would like to thank the following for their help in the preparation of this book: David Gwyn Jones; Roger Andrew Mintern; Mullard Limited; Philips Electronics

MIDWAY JR. HIGH LIBRARY
MENAN, IDAHO 83434